DeltaScience ContentReaders™

Minerals, Rocks and Fossils

Contents

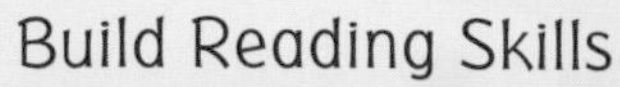

Build Reading Skills

Preview the Book

You read nonfiction books like this one to learn about new ideas. Be sure to look through, or *preview*, the book before you start to read.

First, look at the title, front cover, and table of contents. What do you guess you will read about? Think about what you already know about minerals, rocks, and fossils.

Next, look through the book page by page. Read the headings and the words in bold type. Look at the pictures and captions. Notice that each new part of the book starts with a big photograph. What other special features do you find in the book?

Headings, captions, and other features of nonfiction books are like road signs. They can help you find your way through new information. Now you are ready to read!

What Are Minerals?

MAKE A CONNECTION

Table salt is made from halite. Halite is a mineral. What do you think minerals are?

FIND OUT ABOUT

- materials found in nature called minerals
- properties that can be used to tell minerals apart

VOCABULARY

mineral, p. 4
streak, p. 5
luster, p. 5
hardness, p. 6

About Minerals

A **mineral** is a solid, nonliving material found in nature. We can find minerals in the ground. Minerals make up rocks. Gold, graphite, diamond, and quartz are examples of minerals.

Thousands of different kinds of minerals have been discovered on Earth. But most of the rocks on Earth are made from just a few of those kinds.

Like all materials, minerals are made of tiny particles called *atoms*. The atoms of a mineral are set in a regular, repeating pattern. This pattern is called the mineral's crystal structure.

✓ What is a mineral?

We see and use many minerals every day. ▼

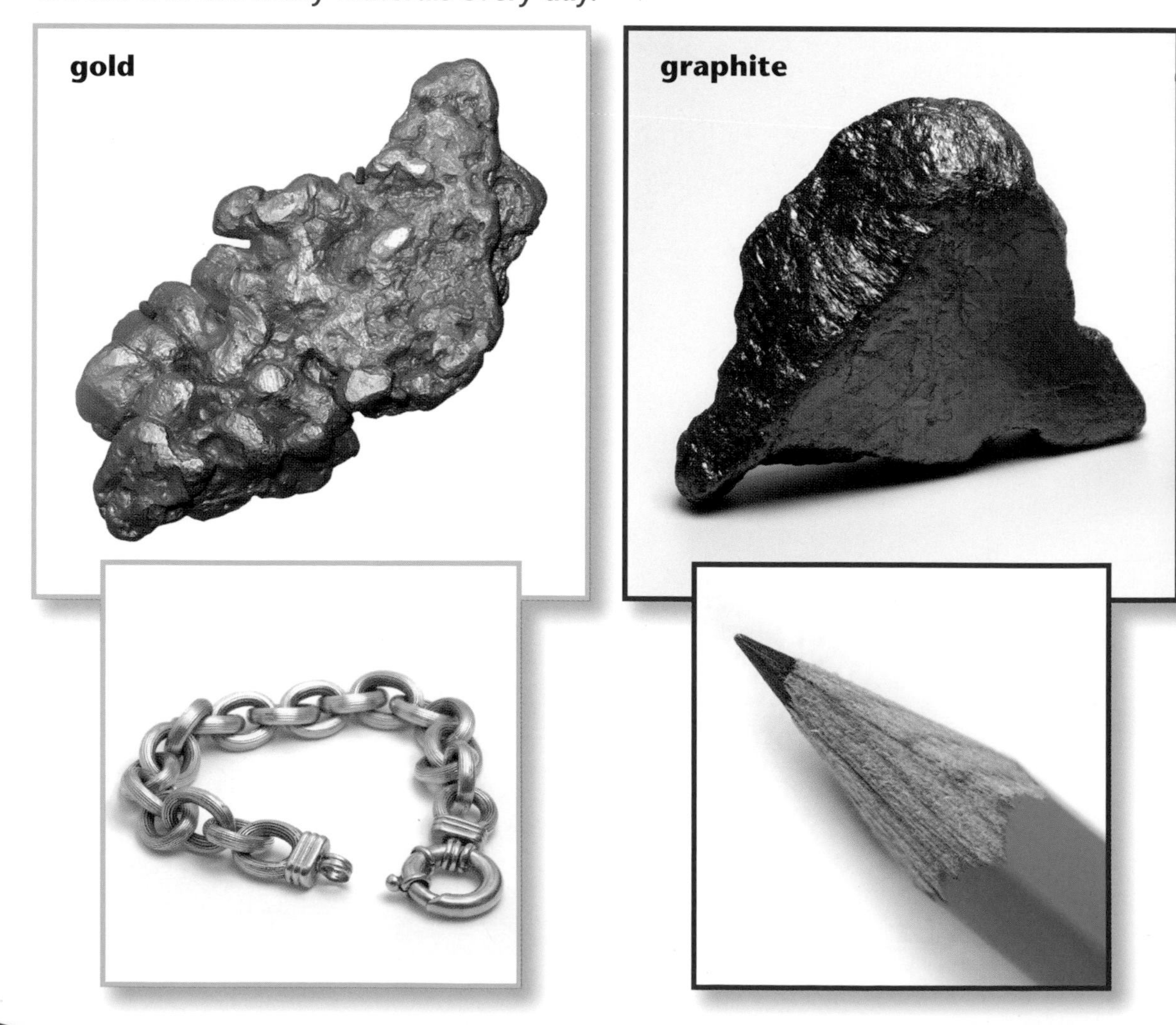

▲ Quartz comes in many colors.

▲ Hematite has a reddish brown streak.

Identifying Minerals

Each mineral has its own physical properties. A physical property is something that can be observed or measured. We can use the following properties to tell minerals apart.

Color Some minerals, such as gold, come in only one color. So color can be helpful in identifying them. But other minerals, such as quartz, come in many colors.

Streak The color of a mineral's powdered form is called **streak**. A mineral's streak is always the same color. To see a mineral's streak, you can rub the mineral on a special tile. The tile is called a streak plate.

Luster The way light bounces off a mineral is called **luster**. Some minerals, such as gold, have a metallic luster. They shine like polished metal. Other minerals, such as talc, have a nonmetallic luster. Some nonmetallic minerals look dull, pearly, waxy, or glassy.

Mica breaks into pieces that have smooth, flat sides. It has cleavage. ▶

Cleavage and Fracture Minerals that break into pieces that have smooth, flat sides have *cleavage*. Minerals that break into pieces that have uneven or curved sides have *fracture*.

Hardness Some minerals are harder than others. **Hardness** is how easy or difficult it is to scratch a mineral. A harder mineral or object will scratch a softer one. The Mohs hardness scale can be used to rank a mineral's hardness from 1 to 10.

Other Properties Some minerals have other special properties. For example, the mineral magnetite is magnetic. It attracts certain metals. The mineral copper is a good conductor of electricity. Electric charge passes through it easily.

Name five properties of minerals.

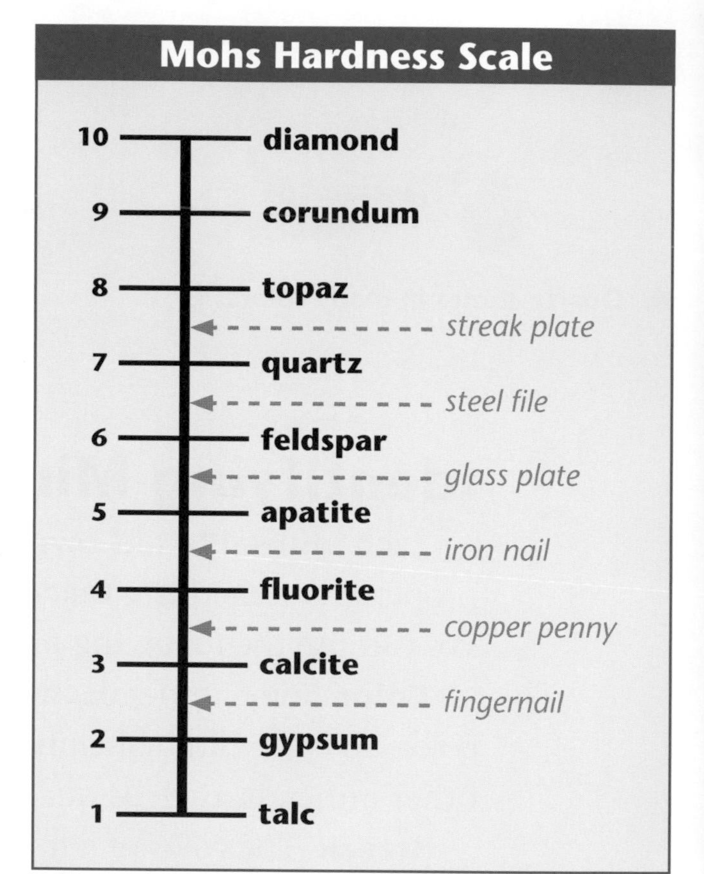

▲ The Mohs hardness scale shows some minerals and objects in order from 1 (softest) to 10 (hardest).

◀ Magnetite is magnetic.

Properties of Six Common Minerals

Mineral	Color(s)	Streak	Luster	Cleavage/ Fracture	Mohs Hardness
calcite	colorless, white	white	nonmetallic	cleavage	3
feldspar	white, pink, gray	white	nonmetallic	cleavage	6
hematite	brown, red, gray, silver	reddish brown	metallic or nonmetallic	fracture	5.5 to 6.5
hornblende	dark green to black	white	nonmetallic	cleavage	5 to 6
mica	black, silver-white, brown	white	nonmetallic	cleavage	2 to 3
quartz	colorless, white, rose, smoky, purple, brown, green, yellow	white	nonmetallic	fracture	7

A table like this can help you identify a mineral. These are six common minerals that make up rocks.

REFLECT ON READING

You previewed pictures, captions, and other book features before reading. Which book features were most helpful when you read about minerals? How did they help you?

APPLY SCIENCE CONCEPTS

Suppose that you find a white, nonmetallic mineral. The mineral has fracture and a white streak. Use the table on this page. What kind of mineral might this be? Tell why.

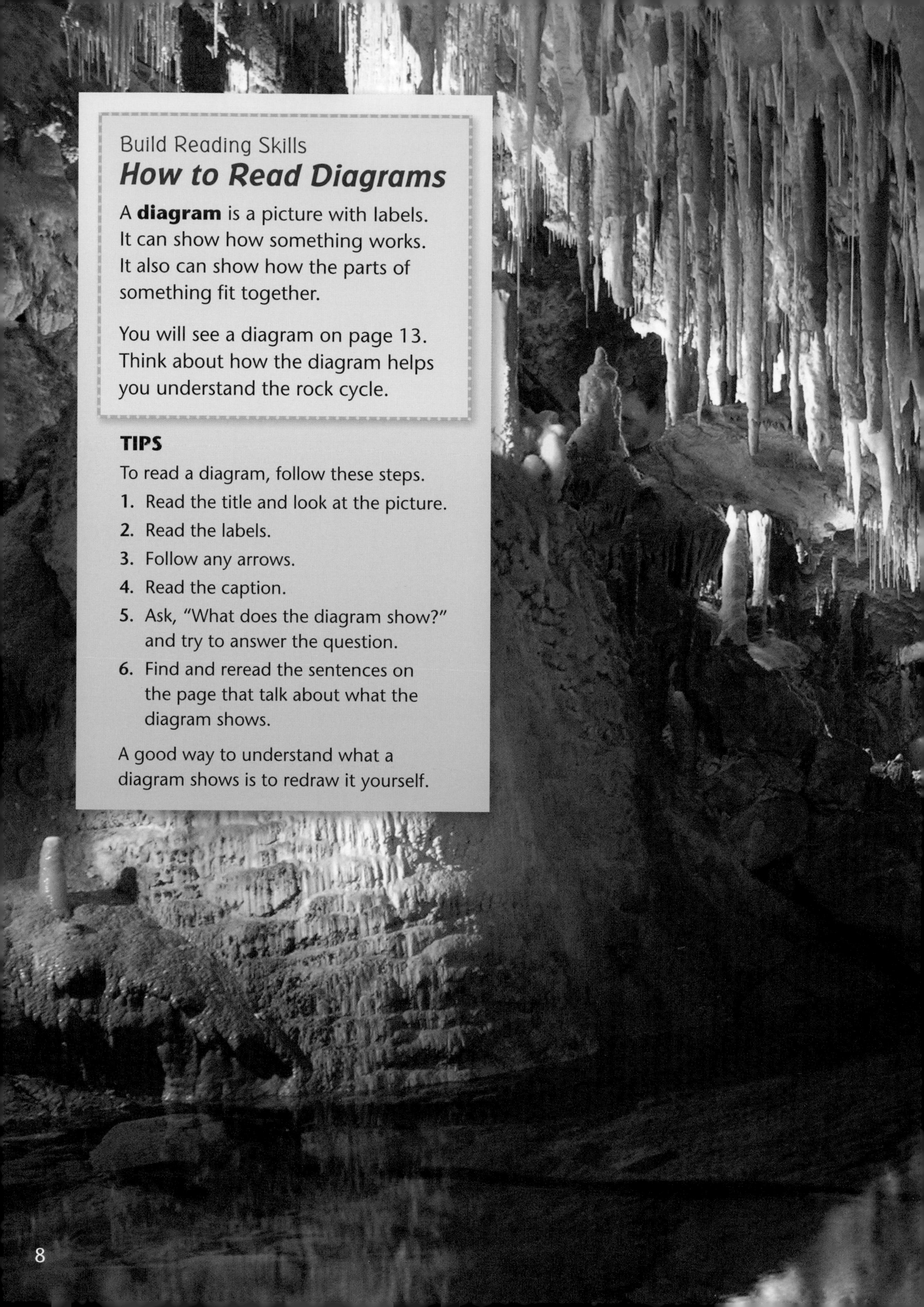

Build Reading Skills

How to Read Diagrams

A **diagram** is a picture with labels. It can show how something works. It also can show how the parts of something fit together.

You will see a diagram on page 13. Think about how the diagram helps you understand the rock cycle.

TIPS

To read a diagram, follow these steps.

1. Read the title and look at the picture.
2. Read the labels.
3. Follow any arrows.
4. Read the caption.
5. Ask, "What does the diagram show?" and try to answer the question.
6. Find and reread the sentences on the page that talk about what the diagram shows.

A good way to understand what a diagram shows is to redraw it yourself.

What Are Rocks?

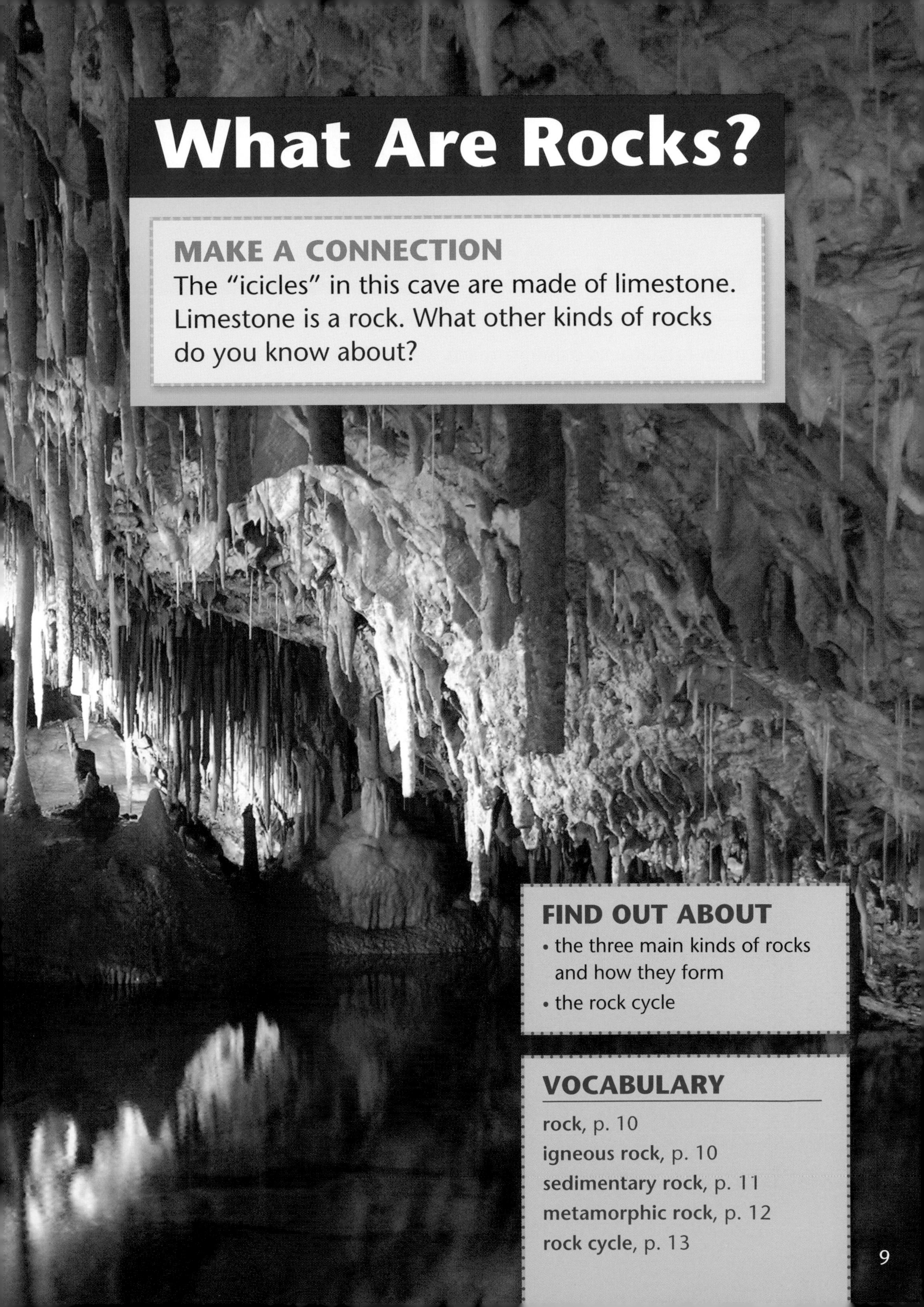

MAKE A CONNECTION

The "icicles" in this cave are made of limestone. Limestone is a rock. What other kinds of rocks do you know about?

FIND OUT ABOUT

- the three main kinds of rocks and how they form
- the rock cycle

VOCABULARY

rock, p. 10
igneous rock, p. 10
sedimentary rock, p. 11
metamorphic rock, p. 12
rock cycle, p. 13

As this lava cools, it will harden into igneous rock. ▶

Kinds of Rocks

A **rock** is a natural solid that is made of one or more minerals. Rocks come in many sizes, shapes, colors, and textures. The properties of rocks can give us clues about how they formed. We can sort rocks into three main groups.

Igneous Rocks

Igneous rocks form when very hot, melted rock material cools and hardens. Some igneous rocks, such as granite, form underground from magma. *Magma* is melted rock material that is below Earth's surface. Underground, magma cools and hardens slowly. This can form rocks that have large mineral crystals.

Other igneous rocks, such as basalt, form on Earth's surface from lava. *Lava* is magma that has reached Earth's surface. Above ground, lava cools and hardens quickly. This can form rocks that have small mineral crystals.

Examples of Igneous Rocks

Granite forms when magma underground cools and hardens.

Basalt forms when lava on Earth's surface cools and hardens.

Sedimentary Rocks

Sedimentary rocks form from minerals and rocks that have been broken down into smaller pieces. The smaller pieces are called *sediment.* Sand is an example of sediment. It can take millions of years for sedimentary rocks to form.

Most sedimentary rocks, such as shale, form in five main steps.

1. Sediment is made when minerals and rocks are broken down, or weathered. Water, temperature changes, wind, and even plant roots can cause weathering.
2. The sediment is moved, or eroded. Wind, moving water, ice, and gravity can move sediment.
3. The sediment is dropped, or deposited, in a new place. Newer sediment is deposited on top of older sediment.
4. The top layers of sediment press down, or compact, the bottom layers. The layers are squeezed together.
5. Water deposits other minerals in the sediment. The minerals join the sediment together like cement.

Other kinds of sedimentary rocks form from minerals left behind when a sea or a lake dries up. Some kinds of limestone and rock salt form this way.

Still other sedimentary rocks form from materials that were once part of living things. Coal is an example.

Examples of Sedimentary Rocks

Shale forms when layers of sediment are squeezed together and cemented.

Some kinds of limestone form when water with minerals in it dries up.

Bituminous coal forms when dead plants are buried by sediment and pressed together.

Metamorphic Rocks

Metamorphic rocks form when rocks are changed by heat and pressure. Most metamorphic rocks form deep underground. There, rock is pressed and squeezed by rock above and around it. All that pressure makes the temperature go up. The rock does not melt. But the way the rock looks changes. The minerals in the rock also can change into different minerals.

Metamorphic rocks can form from any kind of rock. The original rock is called the parent rock.

Slate is a metamorphic rock. It forms when the sedimentary rock shale is changed by heat and pressure.

Gneiss is another metamorphic rock. It can form when the igneous rock granite is changed by heat and pressure.

 What is a rock? What are the three main kinds of rocks?

Gneiss is a metamorphic rock. It often looks striped. ▼

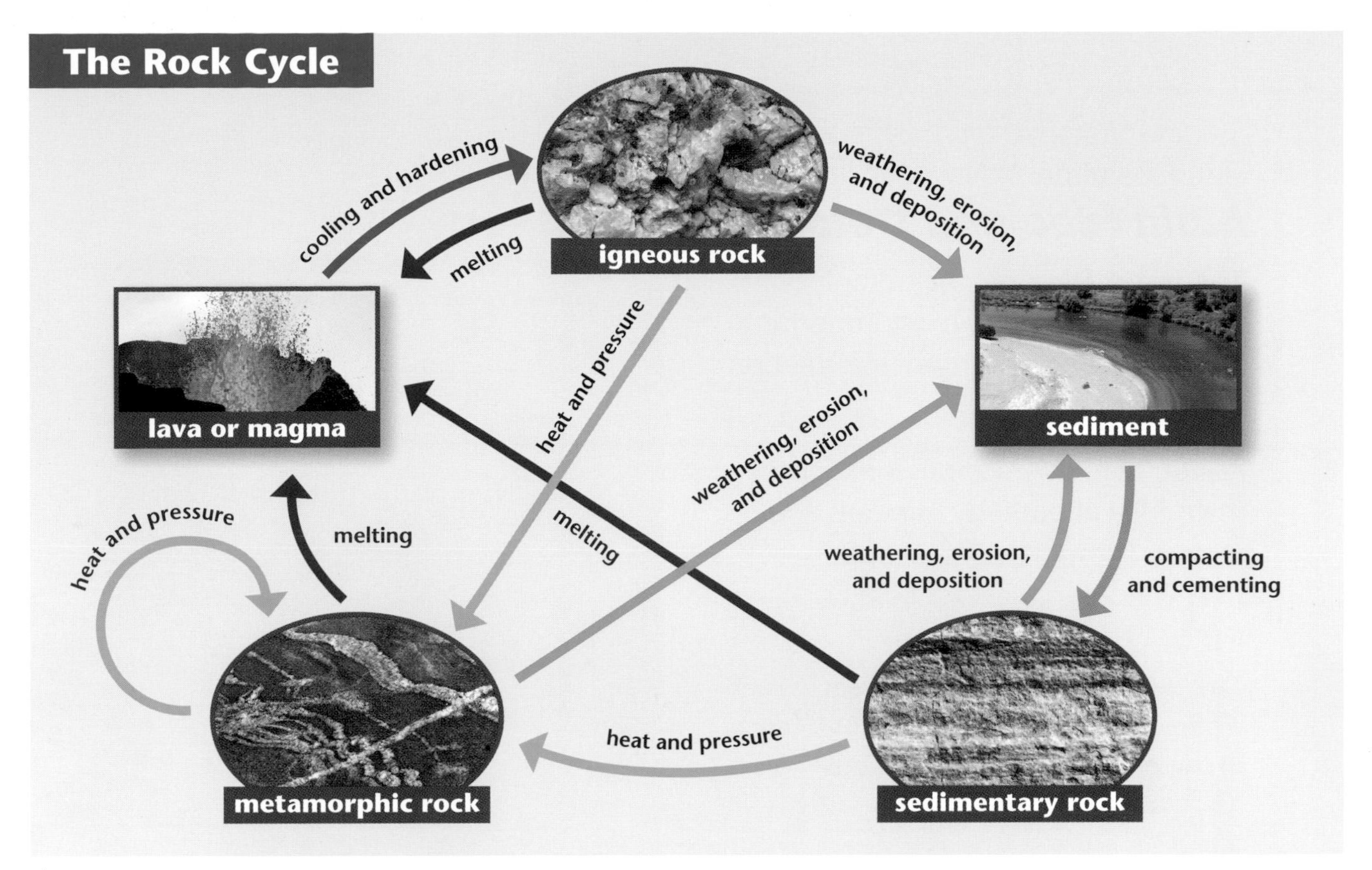

▲ The rock cycle includes all the ways rocks can be changed.

The Rock Cycle

Rocks are always being formed, broken down, and formed again. All the ways rocks can be changed make up the **rock cycle**.

Any kind of rock can be changed to any of the other kinds. For example, look at the igneous rock in the diagram. Weathering, erosion, and deposition change igneous rock to sediment. Compacting and cementing change sediment to sedimentary rock.

Heat and pressure also can change rocks. So can melting, cooling, and hardening.

 What is the rock cycle?

REFLECT ON READING

Look again at the diagram of the rock cycle on this page. Choose one kind of rock. Tell a partner how your rock can be changed into another kind of rock.

APPLY SCIENCE CONCEPTS

Invite someone to share their rock collection with the class. Look closely at some rocks. Use books from the library to learn if they are igneous, sedimentary, or metamorphic.

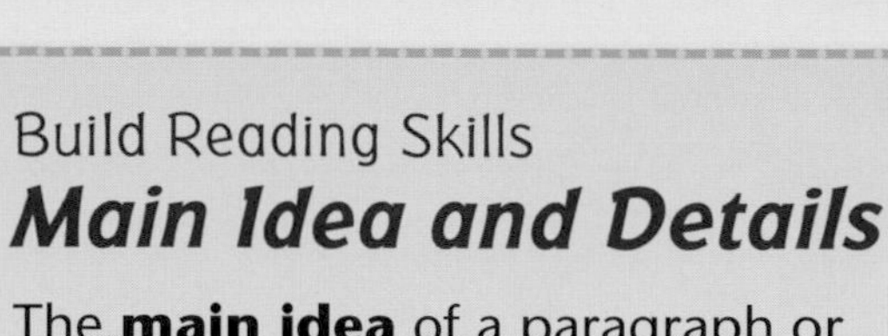

Build Reading Skills

Main Idea and Details

The **main idea** of a paragraph or part of a book is the most important point. **Details** give more information about the main idea.

As you read this section, look for the main idea about why minerals and rocks are important to people.

TIPS

The topic sentence tells the main idea of a paragraph. It is often the first sentence in the paragraph. To find the main idea, ask, "What is this paragraph mostly about?"

Details may answer Who, What, When, Where, Why, and How questions about the main idea. Details can be

- examples
- descriptions
- reasons
- other facts

A concept web can help you keep track of the main idea and details.

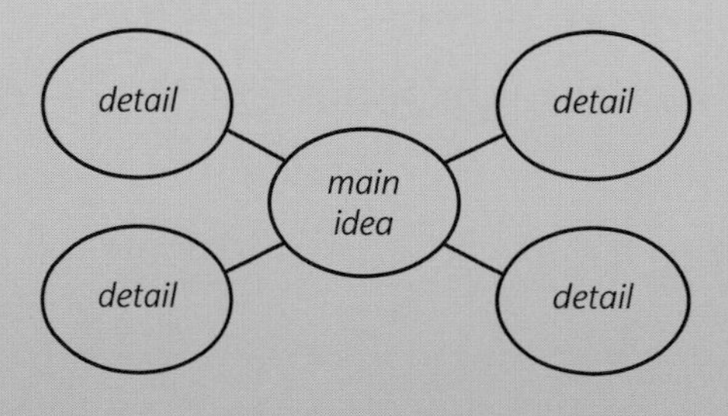

How Do We Use Minerals and Rocks?

MAKE A CONNECTION

This is a copper mine. Copper is a metal. It is used to make wires and pipes. Can you name some other materials that people mine and use?

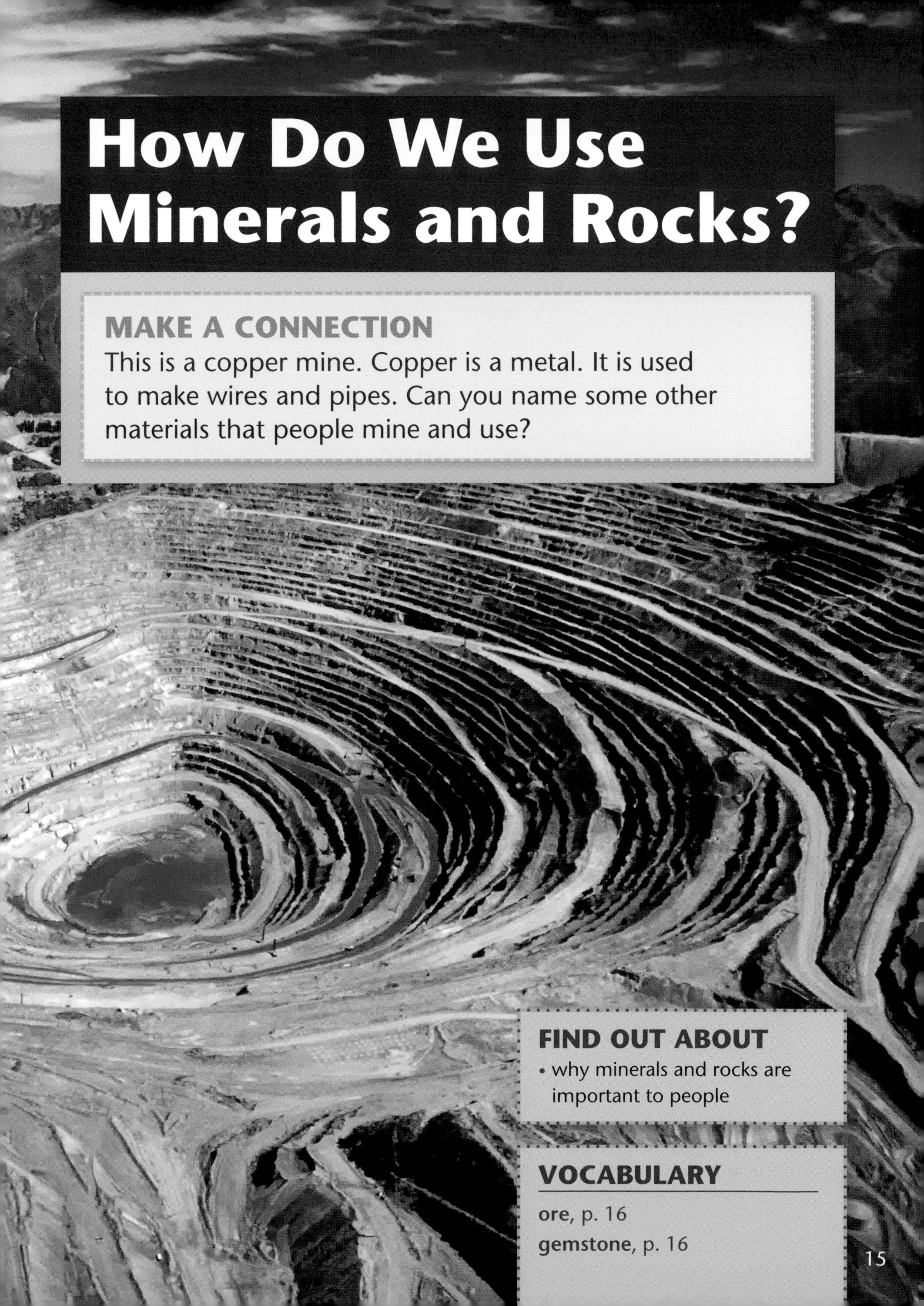

FIND OUT ABOUT

- why minerals and rocks are important to people

VOCABULARY

ore, p. 16
gemstone, p. 16

We get valuable metals from mineral ores. ▶

Some Important Mineral Ores

Mineral Ore	Metal	Some Uses
bauxite	aluminum	building materials, cars, airplanes, aluminum foil, cans, pots and pans
chalcopyrite	copper	electrical wires, pipes for plumbing and heating
galena	lead	batteries for cars and trucks, welding
hematite	iron	steel for buildings, bridges, cars, ships, large appliances

Minerals and Rocks as Resources

Resources are things that we use to meet our needs. Minerals and rocks are important resources. For example, weathered minerals and rocks make up a large part of soil. We use soil to grow plants for food.

Most minerals and rocks take a long time to form. So they cannot be replaced easily if they are used up. They are *nonrenewable* resources.

People remove, or mine, minerals from the ground and use them in different ways. Some minerals are mined for the valuable materials, such as metals, that are in them. These minerals are called **ores**.

Other minerals are mined to be cut and polished for jewelry. These minerals are called **gemstones**. Diamonds, emeralds, and rubies are gemstones.

Minerals called gemstones are rough and dull at first. After cutting and polishing, they can look like this. ▶

◀ **These are piles of the mineral called halite. Halite is used for table salt, paper, and soaps.**

We use minerals, or things made with them, every day. Table salt is made from the mineral halite. Halite is also used to make paper and soaps. The mineral quartz is used to make glass and jewelry. Quartz is also used in parts for computers and cell phones.

Rocks are also resources. For example, we often use rocks for building materials. Some sedimentary rocks are used to make cement, bricks, and drywall. Slate, a metamorphic rock, is sometimes used for roofs and floors. Marble, another metamorphic rock, is often used to make statues.

Bituminous coal is a sedimentary rock that we mine and use for fuel. Many power plants burn coal to run generators that make electricity.

 Why are minerals and rocks called resources?

REFLECT ON READING

Make a concept web like the one on page 14. Use the web to keep track of why minerals and rocks are important to people. Put the main idea in the middle. Then add at least four details, such as examples.

APPLY SCIENCE CONCEPTS

Choose an ore from the table on page 16, the library, or the Internet. Research your ore. Where is it found? Look for interesting facts. Tell the class what you learn.

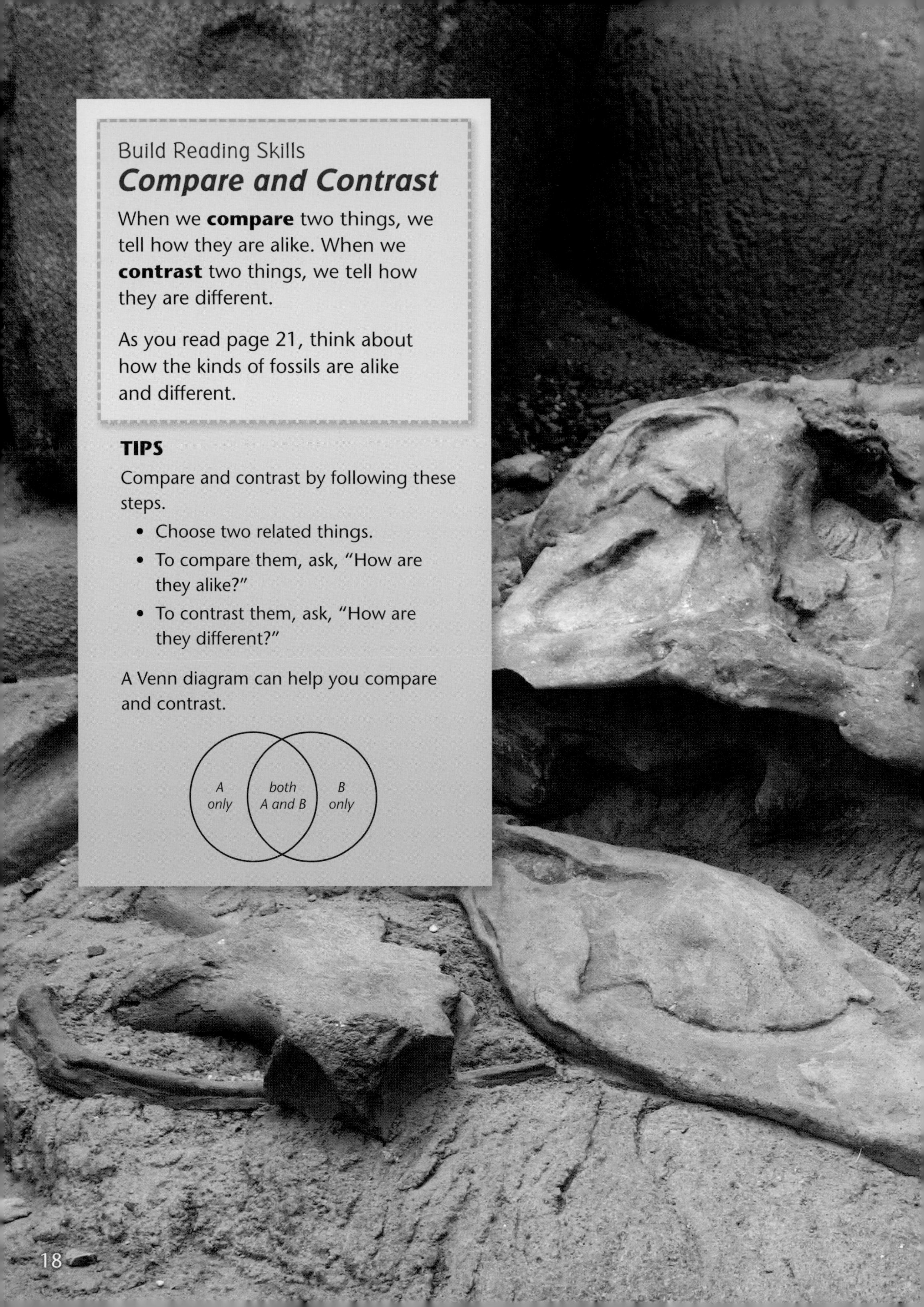

Build Reading Skills

Compare and Contrast

When we **compare** two things, we tell how they are alike. When we **contrast** two things, we tell how they are different.

As you read page 21, think about how the kinds of fossils are alike and different.

TIPS

Compare and contrast by following these steps.

- Choose two related things.
- To compare them, ask, "How are they alike?"
- To contrast them, ask, "How are they different?"

A Venn diagram can help you compare and contrast.

What Are Fossils?

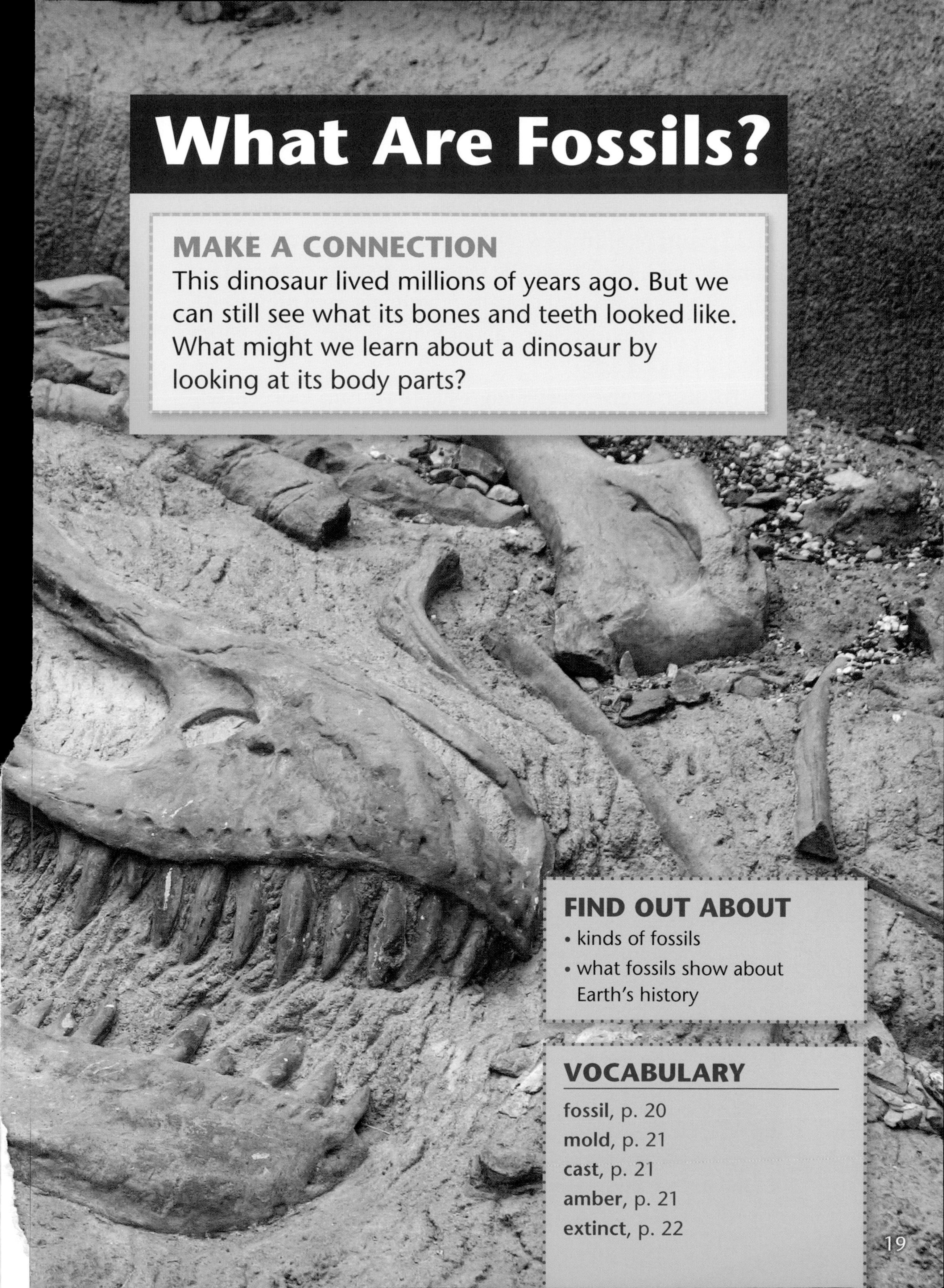

MAKE A CONNECTION

This dinosaur lived millions of years ago. But we can still see what its bones and teeth looked like. What might we learn about a dinosaur by looking at its body parts?

FIND OUT ABOUT

- kinds of fossils
- what fossils show about Earth's history

VOCABULARY

fossil, p. 20
mold, p. 21
cast, p. 21
amber, p. 21
extinct, p. 22

About Fossils

Fossils are the remains or traces of living things from long ago. Some fossils are hundreds of millions of years old.

Most fossils are found in sedimentary rock. Sometimes dead plants or animals were covered by sediment such as sand or mud. Over time, the sediment became sedimentary rock. The remains of the plants or animals were preserved. Fossils help scientists learn about living things, or *organisms,* that lived long ago.

Fossils can give clues about what organisms looked like. Fossils of the hard parts of organisms are the most common. Bones, teeth, and shells are examples.

Fossils also can give clues about how organisms acted. Tracks can show how animals moved. Teeth and droppings can show what animals ate.

▲ **This fish fossil was found in layers of sedimentary rock.**

▲ **These fossils both formed from the body of a trilobite. Trilobites were animals. They lived more than 250 million years ago.**

Scientists have discovered many kinds of fossils.

- A fossil imprint forms when an organism makes marks in mud. Then the mud hardens into rock. Animal tracks and leaf prints are examples.
- A fossil **mold** forms when a dead organism is buried in sediment. As the sediment turns to rock, groundwater dissolves the organism's body. A hollow space, or mold, in the shape of the organism is left in the rock. Groundwater sometimes deposits minerals in a mold. The minerals harden into a model, or fossil **cast**, of the organism.
- A petrified fossil can form when groundwater seeps into a dead organism. Minerals in the water slowly fill in or replace parts of the organism. Petrified wood and bones are examples.
- Fossils of whole organisms are sometimes found in ice, tar, or amber. **Amber** is hardened tree sap.

✓ Tell what fossils are. Where are most fossils found?

Fossils and Earth's History

All the fossils on Earth make up the fossil record. This record can help us learn about the history of life on Earth.

The fossil record shows that most kinds of organisms that lived in the past are not alive today. They are **extinct**. Dinosaurs are a large group of extinct animals. Many other kinds of animals and plants are also extinct.

The fossil record shows us ways extinct and living organisms are alike and different. Scientists compare fossils to other fossils. They also compare fossils to organisms that are alive today. Scientists have studied dinosaur fossils this way. They have learned that birds and crocodiles are probably related to dinosaurs.

The fossils in this museum are just a small part of the whole fossil record. ▼

The fossil record shows the conditions surrounding organisms in the past. We have learned from fossils that Earth's environments have changed over time. For example, fossils of plants that would have lived well in humid places have been found in deserts. So those deserts were once much wetter than they are now.

The fossil record also shows that Earth's land has changed over time. Scientists found the same kinds of fossils on different continents. Those continents are now separated by oceans. The organisms that left the fossils could not have swum that far. Scientists also studied rocks and the shapes of the continents. They concluded that Earth's continents were once joined together. Over time, the continents moved to where they are now.

 What is the fossil record?

Earth's Moving Continents

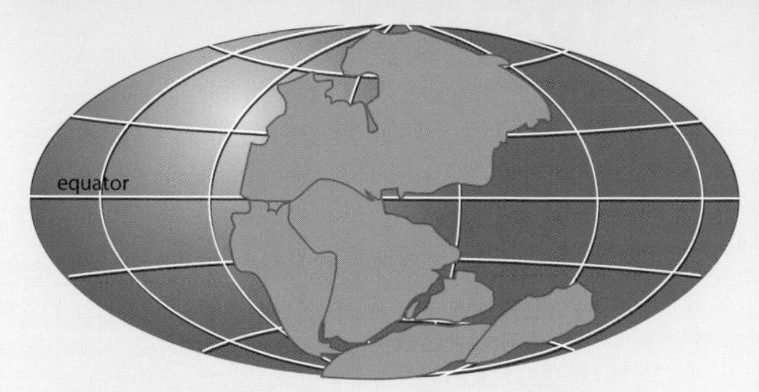

225 million years ago

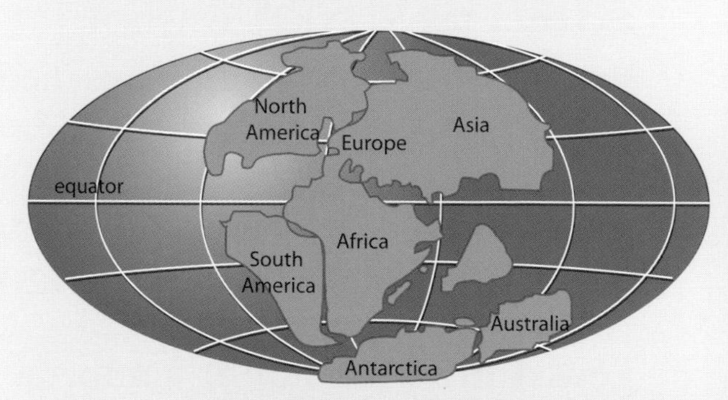

135 million years ago

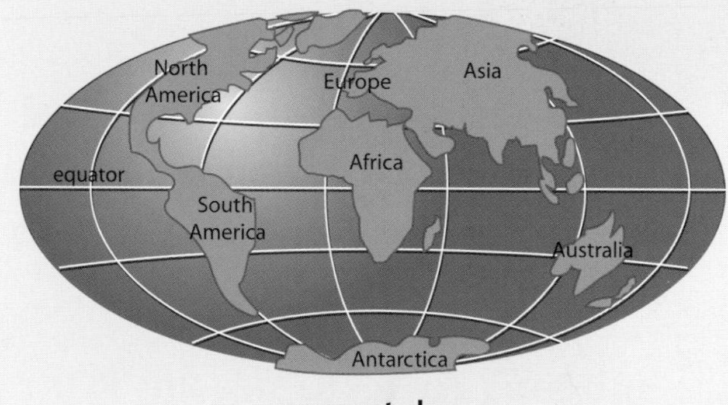

present day

▲ The continents have moved over time. They are still moving today. Studying fossils and rocks helped scientists learn about this.

REFLECT ON READING

Make a Venn diagram like the one on page 18. Choose two of the kinds of fossils from page 21. Use the Venn diagram to show how they are alike and different.

APPLY SCIENCE CONCEPTS

Suppose a scientist finds fossils of ocean animals on a mountain. What might he or she think about the history of that place? Why? Write in your science notebook.

Glossary

amber (AM-buhr) hardened tree sap **(21)**

cast (KAST) a kind of fossil that forms when a fossil mold is filled with minerals **(21)**

extinct (ek-STINKT) when all the living things of one kind have died out **(22)**

fossil (FOS-uhl) a body part or a trace of a living thing from long ago **(20)**

gemstone (JEM-stohn) a mineral that can be cut and polished for use in jewelry **(16)**

hardness (HARD-nis) a physical property that tells how hard a mineral is and how easy or difficult it is to scratch the mineral **(6)**

igneous rock (IG-nee-uhs ROK) a kind of rock that forms when hot, melted rock material cools and hardens **(10)**

luster (LUHS-tuhr) a physical property that tells how light bounces off a mineral **(5)**

metamorphic rock (met-uh-MOR-fik ROK) a kind of rock that forms when rocks are changed by heat and pressure **(12)**

mineral (MIN-uhr-uhl) a solid, nonliving material found in nature; minerals are the building blocks of rocks **(4)**

mold (MOHLD) a kind of fossil that is a hollow space in rock left by the body parts of a living thing **(21)**

ore (OR) a mineral that is mined for the valuable materials, such as metals, that are in it **(16)**

rock (ROK) a natural solid that is made of one or more minerals **(10)**

rock cycle (ROK SYE-kuhl) all the ways rocks and rock material can be changed from one kind to another **(13)**

sedimentary rock (sed-uh-MEN-tuh-ree ROK) a kind of rock that forms when layers of sediment are pressed together and naturally cemented **(11)**

streak (STREEK) a physical property that tells the color of a mineral's powdered form **(5)**